Roku User Manual Guide: Private Channels List, Tips & Tricks

By Shelby Johnson

http://techmediasource.com

Contents

Introduction: Why Choose Roku

When it comes to live streaming your favorite programs, the Roku streaming device has all of the answers. With several different types of media streaming "boxes" to choose from, you can turn your entertainment wants and needs into a vivid reality on your television screen.

So, why choose the Roku over a Chromecast or Apple TV? Simple. First, there are levels of purchase that allow for individual enjoyment. With even the lowest price point device, starting at $49.99, you can add the box to virtually any television, while gaining access to 1000+ entertainment channels. With channel shortcut buttons, built-in wireless and one stop search included with the original, each upgraded box only adds to the allure of Roku and its sensational offerings.

For the same price you can purchase Chromecast, as it is available for $35. Although the entertainment offerings as similar to the initial Roku setup, the device is very Google-centric (since it is their design and operation), so all things Google will apply, like a Google Play account and searches based on their criteria and return algorithms. In addition, using a Chromecast requires streaming media to the device using a laptop or mobile device, whereas a Roku does the streaming for you.

You cannot purchase an Apple TV for the same price, as it begins and ends at $99, which is the same price as the Roku 3 model that provides dual band wireless, motion controls for games and an Ethernet, USB and microSD slot located directly on the device for easy file transfers directly to your screen.

As of right now, the Roku offers the most variety when it comes to streaming content channels including video, music, movies, live TV, news, sports, weather and much more. For the price point, it makes a lot of sense for those looking to save money on their cable bill while still enjoying great content on their TV.

For more info on enjoying TV without an expensive cable bill, see my other book, How to Get Rid of Cable TV and Save Money.

Types of Rokus

There are at least four types of Roku devices that are designed and tailored to meet the needs of individuals and their entertainment desires. Each provides its very own capabilities while, as you would expect, each tier brings a little more to the entertainment table, for just a little more cash.

The Roku LT

Price: $49.99

The Roku LT 2013 model is available to the masses for a mere $49.99, and is compatible with virtually every television on the market today. It provides access to 1000+ channels of entertainment, built-in wireless, and one-stop search across multiple channels and channel shortcut buttons. The entertainment offerings from the original Roku stream at 780p HD Video.

Roku 1 (HD)

Price: $59.99

The Roku 1, or HD Roku, can be purchased (where applicable) for the same price as the original, beginning at $49.99 on the company' website. With the upgraded version you will receive everything that comes with The Original, including access to 1000+ channels of entertainment, built-in wireless, and one-stop search across multiple channels and channel shortcut buttons. However, the content streams in 1080p HD video.

Roku 2

Price: $79.99

With the Roku 2 device, you get everything you get with the first two devices, including access to 1000+ channels of entertainment, built-in wireless, one-stop search across multiple channels, channel shortcut buttons, 1080p HD video and add a remote with headphone jack, and a dual-band wireless connection.

Roku 3

Price: $99.99

Finally, the Roku 3. With a price point matching Apple TV, the Roku 3 dominates the offerings, starting with the standard delivery of the first three versions, including access to 1000+ channels of entertainment, built-in wireless, one-stop search across multiple channels, 1080p HD video, remote with headphone jack, and a dual-band wireless connection. Now add motion control for games, a processor that is five times faster than its predecessors, an Ethernet, USB and microSD slot and an updated version of both YouTube and Netflix apps, including a Send to TV feature.

3400M Streaming Stick

The Roku 3400M Streaming Stick allows you to forgo the electronic box altogether, and simply plug the dongle into your Roku-ready television or Blu-ray player to receive access to the thousands of programs available through their service. The price for the streaming stick has normally been below $80.

Roku: What's in the Box?

Depending on which version of the Roku you purchase, there are different components in the box it comes in. Each of those offerings are broken down below, along with the necessities for operating each for quick reference.

The Original Roku: In the Box

- Roku LT Streaming Player
- Roku Standard Remote Control with Channel Shortcut Buttons
- Two AAA Batteries
- Composite A/V cable (red/white/yellow)
- Power Adapter
- Get Started Guide

What You Need to Operate the Original

- HDTV or Standard Definition TV
- High-speed Internet (like DSL or Cable)
- Wireless Router
- HDMI Cable (for High-definition Video)

Roku 1: In the Box

- Roku 1 Streaming Player
- Roku Standard Remote Control with Channel Shortcut Buttons
- Two AAA Batteries
- Composite A/V Cable (red/white/yellow)
- Power Adapter
- Get Started Guide

What You Need to Operate the Roku 1

- HDTV or Standard Definition TV
- High-speed Internet (like DSL or Cable)
- Wireless Router
- HDMI Cable (for High-definition Video)

Roku 2: In the Box

- Roku 2 Streaming Player
- Roku 2 Enhanced Remote Control with Channel Shortcut Buttons and Headphone Jack
- Two AAA Batteries
- Composite A/V Cable (red/white/yellow)
- Power Adapter
- Get Started Guide

What You Need to Operate the Roku 2

- HDTV or Standard Definition TV
- High-speed Internet (like DSL or Cable)
- Wireless Router
- HDMI Cable (for High-definition Video)

Roku 3: In the Box

- Roku 3 Streaming Player
- Roku 3 Enhanced Game Remote and Headphone Jack
- In-ear Headphones
- Two AA Batteries
- Power Adapter
- Get Started Guide
- Free Full Angry Birds Space Game

What You Need to Operate the Roku 3

- HDTV with HDMI Input
- HDMI Cable

- High-speed Internet (like DSL or cable – Roku recommends a minimum of 1.5 Mbps for standard definition and 3.0 Mbps for HD content.)
- Wired or Wireless Router

Roku Streaming Stick: In the Box

- Roku Streaming Stick
- AA Batteries

What You Need to Operate the Roku Streaming Stick

- Broadband Internet (min 1.5 Mbps)
- WiFi (802.11 a/b/g/n) Wireless Network
- High Definition TV
- HDMI Cable
- Roku Account (free)
- Internet Browser to Link Player or Channel

Setting up Roku

Setting up your Roku will completely depend on the type of streaming box (or stick) you purchase, and the quick start guide that comes with each version will show you exactly what to do. You'll want to make sure you have your Internet connection info ready including any login or password info necessary to connect your Roku to your particular router. You will also need any other subscription account information ready when you set up your Roku such as Netflix, Hulu, or other subscription services.

Initial Setup

So you've unboxed your new gadget and now it's time to stream some media. The Roku media devices are rather simple to set up and get going, but here's a guided set of instructions for those who may not have an original start guide with their devices. These instructions are for the Roku 3, the newest of the devices, so they may vary slightly for other models.

1. Connect one end of HDMI cable to an input area on your HDTV monitor.
2. Connect another end of HDMI cable to the input on your Roku device.
3. Make sure batteries are inserted into your Roku remote in the correct positions.

4. Plug Roku device into included power cable and plug other end into a socket. The Roku will take a few moments to boot up.
5. Select your preferred language.
6. Assist the Roku in finding your wireless network (Wi-Fi). It will scan for all available networks. Select your network from the choices shown.
   Note: *You may be required to manually enter information for your network such as any special login information.*
   Once your network has been found, Roku will set up the connection between your Wireless network and the Roku device.
7. Press "Continue" on the next pop-up to get the latest software update for your device. You should see a Progress box displaying the percent of the update completed so far. Once the software update has been installed, Roku will automatically restart to complete the process.

Activate Roku with Online Account

If you've already got an online account set up with Roku activate it using the following steps:

1. Go to https://owner.roku.com/Login.
2. Enter your log-in credentials, then scroll down to "My linked devices" and click on the "Link a Device" button.

OR

If this is your first time with Roku activate it by completing the following:

1. Go to roku.com on a PC or MAC computer.
2. Enter the code displayed on your TV screen from your Roku into the box on the Roku.com/link site. (It should be a 6-

character code which is a combination of letters and numbers.)

The next screen pre-selected content channels that will be added to your Roku. These include popular channels such as Netflix, Redbox Instant, Blockbuster on Demand, and YouTube.*
*for Roku 3 only as of this publication's date.

You can go through the webpage and click on "+ Add Channel" on others you may want to add to your Roku. You can also click on any of the already "Selected" buttons to de-select those channels so they won't be added to your device.

Note: *Keep in mind that while many of the channels are free, others will involve a cost which should show up below them. There's more about channels later in this guide.*

Once you have selected the channels on the webpage that you want to add to your Roku, click "Continue" at the bottom of the page. Your Roku device will now be linked with your online account, and the channels you have chosen will be added to the device with an update showing the status on your TV screen.

Navigating the Roku Remote

The Roku remote (Roku 2 below) is lightweight and handy to use, so it's important to understand how to find your way around the device screens using it. It is also helpful to install the mobile apps for Roku on your phone or tablet to make navigation even easier. More on that later on in this guide.

Left Arrow – navigates you back to the previous screen or page you were on with the Roku.

Home Button – returns you to the Roku's home screen which includes your channels and settings.

Directional buttons – In the center of your remote will be several purple directional buttons. These are used to move left, right, up and down among choices or options on your screen. For example, use this to navigate through your channels or to select a movie to watch.

OK button (make selection) – Depending on your version of Roku remote, you will find a purple "OK" button either smack dab in the middle of your directional arrows or below it. Use this button to select the option, channel, or other item you want to go to. The "OK" button is also used to confirm choices on the screen.

Back/Play/Pause/Forward – On the two remotes pictured above are a variety of different navigational options including a back/rewind button, a play/pause button and a forward button. Use these to move back or ahead on various content, or to play or pause the content.

Jump back button (on certain remotes) – A curved arrow pointing towards the left and down represents the ability to move back on content in increments, such as 30 seconds backwards. This particular option may not work on every channel.

Asterisk button (options) – This can be used on various screens to bring up options for those screens. You should see "* Options" highlighted in the right-hand corner on your TV screen. This means the particular screen you are on has options to choose from or additional info. If "* Options" is faded, it means there are no additional options to use for that particular channel or screen.

A/B buttons – Certain Roku remotes will also have a green A button and a blue B button at the bottom area of the remote. These are used when playing certain games on your Roku device.

Dedicated content buttons – Certain Roku remotes also feature buttons that will take you directly to your Netflix, Pandora and Crackle channels on your device. These are Roku partner services associated with the content streaming device.

How to Add Channels

There are a number of channels – hundreds, in fact – available with your Roku box. You will find the readily available options in the Channel Store on your device, or through the Channels section of the Roku website.

However, there are many more channels that exist that are not listed publicly, call private channels. They are absolutely real channels, but go without promotion by Roku. These channels are often created by independent developers and can add to the appeal of your device.

Here's how you add a private channel:

1. Find the channel access code by searching the Roku sites Roku Guide, or Roku Channels. As an example, the channel

access code for NoWhere TV (a streaming TV channel) is H9DWC.
2. Visit owner.roku.com/Add. (You will need to have a Roku account set up already, and it is completely free.
3. Enter your username and password.
4. Click Manage Account > "Add a Private Channel."
5. Insert the channel access code.
6. Confirm you want to add a channel.
7. Click "Return to My Account."

A list of top private channels and their codes is provided later on in this guide.

Free vs. Log-in Channels

There are hundreds of free channels available on Roku, which will allow you to enjoy endless entertainment without the expense of a cable bill. However, there are several log-in channels that require a subscription, including Amazon Prime, Netflix, Hulu Plus, or Pandora. Some of these channels also require that you are a cable or satellite television subscriber with access to the service. Examples might include HBO Go and WatchESPN/ESPN3.

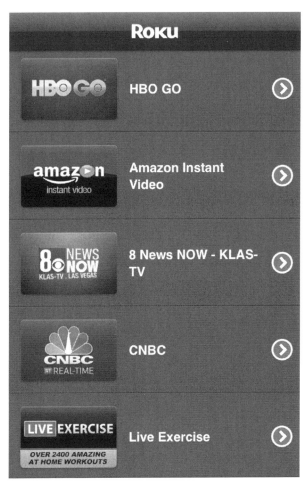

There are also "season pass" style channels, usually for sports fans. Among them are the professional sports leagues such as Major League Baseball and the National Basketball Association. Subscribers pay a season pass fee which allows them to watch multiple games of their choice throughout the team or sport's season.

It is completely up to you which channels you include in your line-up, and you absolutely do not have to subscribe to a single channel that requires payment. Part of the allure of the device is to enjoy it without the costly expense other programming alternatives require, but that is completely up to the individual owner.

Roku Tips and Tricks

Enjoying your Roku is pretty straightforward, but there are a few tips and tricks that can keep your entertainment level thriving, instead of inviting frustration. Here are some of the best ones to kick start your entertainment value.

- If you have a Roku player with a USB port, you can watch your personal videos on your TV using the Roku USB Media Player channel.
- Adding Plex to your Roku allows you to stream your personal media from a laptop, desktop or media server. Add it here, and download the Plex Media Server to get started.
- You can transform your mobile device into a Roku command center with their official Roku app for Android and iPhone.

Change Roku's Theme

The Roku's standard background theme is dark purple and black color schemes. However, it is possible to change up your theme for a different look and feel. To change the background theme:

1. Go to "Settings" on your Roku.
2. Select "Themes" on your Roku.
3. In the "My themes" area you can choose from among the different themes offered such as the Roku default, Daydream, Nebula, Decaf and Graphene.
4. In the "Custom settings" area you can turn "Featured themes" On or Off. These are new themes that arrive

from time to time on Roku such as those for holiday seasons. A Christmas holiday/winter theme is an example of a featured theme you might see. Turn this option on for Roku to automatically display any featured themes on your device as they arrive.

Change Roku's TV Screensaver

A screensaver can be a helpful way to prevent content from staying frozen up on your television screen. To select a screensaver:

1. Go to "Settings" on your Roku.
2. Select "Screensaver" on your Roku.

You can now select from various available screensavers including a Roku Digital or Analog clock, photos from your Facebook timeline, a bouncing Roku logo, and other options.

In addition, you can go to the "Wait time" option and select how long your Roku must be idle before the screensaver is activated. You can also "disable screensaver" there to prevent one from showing up at all.

Roku iOS and Android App

Roku has stepped up their game and added an App for iOS (Apple) and Android operating systems that allows you to turn your smartphone or tablet into a touchscreen remote as well as taking your Roku's features to another level. Simply download the app from the respective store (App Store or Google Play) that corresponds to your device, and begin streaming your videos, photos or access your channels and shopping outlets like Amazon directly from your electronic device.

This app appears to work very well on several devices. The ease of use of typing in login or search info with the on-screen keyboard is especially nice, not to mention the ease of navigating through channels and content. You can browse the channel store and add new ones, or look through content choices easier on your particular mobile device using this app. One other nice feature is the ability to stream any compatible media you might have on your mobile device such as music or videos.

If there's one thing you do right away, make sure to install the Roku remote app on your particular mobile device. You'll be happy you did!

Here are links for the app at the Amazon, iTunes, and Google Play app stores.

Roku Remote at Amazon's App Store:

http://amzn.to/1dGaZSy

Roku Remote at iTunes App Store:

https://itunes.apple.com/us/app/roku/id482066631?mt=8

Roku Remote at Google Play Store:

https://play.google.com/store/apps/details?id=com.roku.remote&hl=en

Use Facebook on Roku

The Roku includes a special free Facebook channel to let you view the photos, videos, statuses and other info from your Facebook right on your TV screen. Here is how to link up your Facebook account on your Roku.

Connect your Device with Facebook

Please enter the Activation Code from the device in the box below.

Enter Activation Code

Connect

1. First, on your Roku, add the Facebook channel if it isn't already.
2. Next, go to the Facebook channel on your device. A screen will pop up giving you a special code to enter.
3. Go to facebook.com/device and enter the code you were given by Roku (see screenshot above).
4. You will get several different pop-up boxes, to continue the process, including one that asks if it's ok for Roku to access your Facebook feed.
5. Once activation completes, the screen should update on your TV, telling you that your Roku player has been successfully linked to your Facebook account.

Once connected, your options with this channel include the ability to view your own photos and videos on Facebook, news feed photos and videos, and your Facebook friends' photos and videos. You can also go into the settings area on the channel and set up a slideshow of the photos which even includes background music, if available. If you leave the Facebook channel idle on your Roku for a certain amount of time, it may automatically show random photos from your timeline up on your TV screen as a screensaver.

Restricting Channels and Content

A helpful tip especially if you have children who might be using your Roku device, is to set up a PIN security code for your account. This will allow you to choose to restrict the device to not allow content purchases or the adding of new channels without entry of a four-digit code number.

To set up the PIN code:

1. Log in to your Roku owner's account on a PC or MAC.
2. Go to "My Account" area of the page.
3. In the middle of the page you should see "PIN Preference."
4. Click the "Update" button.

You can now set up your 4-digit PIN code and choose what this code will need to be entered for on your Roku device.

In-Ear Headphones Jack on Roku 2 and 3

One of the best features of the Roku 2 and 3 is that there is a headphone jack available with the device. This may not seem like much, but it certainly comes in handy when you are trying to watch something in the much hard to come by privacy of a roommate filled home. The headphones allow you to enjoy your programming of choice, wherever and whenever you would like, without disrupting the person next to you who is not interested in your programming choices. Simply plug the headphones directly into the device, and enjoy the private viewing experience.

Use Roku as an Internet Radio

The Roku has all sorts of great music streaming options, including Pandora and Spotify. These services are nice, but will require subscription fees. Another option is the tunein radio channel. This channel provides local radio stations based on the location of your wireless network including popular genres and police scanner or other information stations. There's also a plethora of music folders that Roku users can browse through to find stations based on their favorite genres. With everything from Christmas music to Classical to Hip Hop and Classics, it's a great way to keep music going in the background for a party or other occasions. See more info at the tunein website to start an account and find great music to play via Roku or other mobile devices.

Gaming on Roku 3

Roku 3 comes ready for gaming, with the Angry Birds Space game preloaded to the device. What's more is that the motion-controller that comes standard with the Roku 3 model allows users to download and play movement games as they see fit, a la the Wii. Keep in mind that many games may be available for a fee, so be cautious before adding any game channels with regards to extra costs.

Favorite Channels on Remote

Locating your favorite channels on the remote makes it a lot easier to enjoy your specific programs at the touch of a button. It also keeps you from flipping through the endless amount of channels that are available, wasting time and energy only to return to the same favorites over and over. Roku 2, Roku 1 and Roku LT (original) remotes have channel shortcut buttons that instantly launch Netflix, M-GO, Amazon Instant Video, or Blockbuster On Demand.

Move Channels for Better Organization

Loc You can easily move around your channels on the "My Channels" screen to make it more organized, based on your personal preferences. For example, you can put all of your channels into alphabetical order, or arrange them based on which channels you tend to use most often.

To move a channel:

1. Go to the "My Channels" screen and hover the selector box over the channel you want to move.
2. Press the asterisk "*" symbol on your Roku remote for "Options."
3. Select the "Move" option on the screen and press "OK."
4. Use the directional arrows on your Roku remote to move the channel icon left, right, up and down, until you have it positioned in the spot you want. Press "OK" when the channel is in your desired location.

Repeat this process for all channels until you have them arranged to your liking.

YouTube

In order to enjoy the official YouTube channel on your Roku, a couple of things must first apply to you: You will need to own a Roku 3 player, and you will need to be located in the U.S., Canada, the U.K., or the Republic of Ireland.

If those things apply, you will have access to an official YouTube channel directly from your Roku device that allows you to enjoy all of your online favorites effortlessly. You can enjoy the "Send to TV" option that allows your smartphone and YouTube to be slung directly to your bigger television screen. It also allows you to keep track of your YouTube subscriptions using your Roku.

One possible trick to enjoy YouTube on other Roku devices is to use the Play On browser and media server to stream YouTube videos to your Roku from your Windows-based computer. Keep in mind this service will cost in terms of an annual or lifetime subscription fee. More info at PlayOn.TV.

Search Help

With the amount of channels available from Roku, you may need a little "search help" from time to time. The good news is, you can search the entire device and its entertainment offerings by title, actor or director. Once you find what you want to watch, all you have to do is select the channel it is on, and start watching!

Watchlists

As you are navigating through the various channels and all of the content available, make sure to use the Watchlist options on channels such as Hulu, Netflix and others. You can easily add content that you think you'll like and then go back to that Watchlist later to see the content you noticed before. Otherwise, it can be tough to remember where you saw which movie or show you wanted to watch!

Plex for Roku

Proponents of the Plex application say that it will make users forget about streaming video services like Hulu and Netflix. That might not be totally true, but Plex can take your own media collection and turn it into a unique and personal service that is as easy to enjoy as one of the streaming services. Not only does it help you catalog your digital media files, it also presents them in an attractive way.

The application also includes some popular web-based streaming services, so you should never run out of something to watch or hear. What's even better is that you can access all of that media on your PC and almost any other device that you can think of.

How to Use Plex on A Roku Device

In order to use the application on your Roku device, you need to set it up on your personal computer and Roku. The initial setup should only take a few minutes.

Plex

★ ★ ★ ★ ★ ⋆ 20357 ratings

All Ages

Plex organizes all of your personal media, wherever you keep it, so you can enjoy it on any device. With Plex, you can easily stream your videos, music, photos and home movies to your Roku using your Plex Media Server (available for free at http://www.plexapp.com).

developer   Plex, Inc.
version   2.8

Install Plex on Your Personal Computer

The first thing that you need to do is install the Plex media center and media server on your computer. After these are installed, you are free to:

- **Add media channels:** You can access streaming content from a variety of different places. These include Khan Academy, HGTV, and even the Food Network. It also includes popular streaming sites like YouTube.
- **Add your own media:** Download any media files that you own into the Plex Media Server. This could include movies and music. The great thing about the Plex is that you do not have to worry about the format because the Media Server automatically codes them in the correct format for different devices. This is really convenient for iPhone and Roku users too.

Install Plex on a Roku

Leave your computer on while you install Plex on your Roku. The two devices will connect to each other, and you will see all of your new options on your TV screen or monitor. Again, you do not need to worry about files in different formats because the application can convert them to work on your devices.

As you watch movies or listen to music, you might notice that your personal computer is using up some CPU and accessing your drives. This is actually because Plex might have to covert files to the correct format. However, the process is transparent to you.

How Much Does Plex Cost?

The next best thing about this handy app for computers, smart phones, and all sorts of devices is that you can download it for free to your computer. However, you can get extra features out of the premium version. This version costs $3.95 a month, and you can pay for a year's subscription for $29.99. Some extra features are the ability to enjoy all of your media in the cloud, so you can access your music and movies from anywhere. Paid accounts also can share their media centers, so friends can share music and movies too.

Roku Troubleshooting

Like all electronics, the Roku is not without its head-scratching moments. There are going to be times when you absolutely need a troubleshooting guide, and it is usually for the three items listed below. We are glad to help!

How to Update Roku Software

As an electronic device, the Roku operates with software that from time to time needs to be updated to provide the best viewer experience available. Although the device will push updates towards you by stating that one is available, and asking if you would like to download it, there are times when you are busy with other entertainment options and skip the download. If you choose to do so, it is no problem to update later. Simply take the following steps.

1. Go to the Home Screen by clicking on the "Home" button on your remote.
2. Select "Settings."
3. Select "Software Update."
4. Select "Check Now."
5. Follow the instructions provided on screen when an update is available.

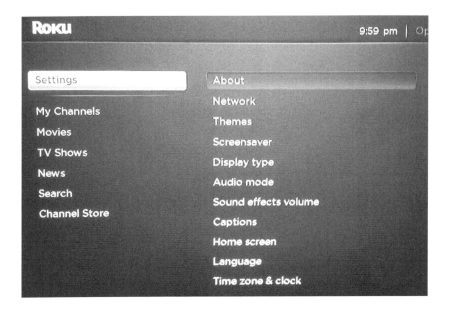

Poor Video or Sound Quality

Poor quality videos simply are not worth watching, so it is important to know that your resolution quality is dependent on two major factors:

- The speed of your broadband connection.
- The strength of the connection between your Roku player and your router (if you are connected wirelessly).

To maximize your broadband connection speed, be sure to discontinue the use of other Internet-connected devices in your home network. Online gaming or other video streaming can use a large portion of your available bandwidth. Finally, to improve your wireless signal strength, try moving your Roku player to a higher position and/or moving it away from other 2.4 GHz interfering sources.

A last resort option might be to try resetting your wired/wireless modem and/or router, but it's important to consult any reference guides included with the devices for performing that method.

Pairing an Unpaired Roku Remote

One issue you may possibly experience is that your Roku remote becomes unpaired and doesn't operate the device. If this is the case, you can try to pair the remote again.

First and foremost, check the batteries in the remote to make sure they are not the issue. Open the battery compartment and remove the batteries, then reinsert them as depicted inside the remote. If the batteries don't appear to be the issue, move to the next steps.

1. Open up the battery compartment and locate the "pairing" button inside.
2. Press and hold down the pairing button for at least 3 seconds. You will see the LED lights blink.
3. Unplug your Roku's power connection and then plug it back. This will power cycle your device.

As the Roku goes through its initial startup process, your remote should be paired with the Roku.

Activating Facebook Channel Issue

As previously mentioned in this guide, you can view your Facebook timeline feed including status updates, photos, videos and other content over your Roku on your television. To do so, you'll need to link your Facebook account with the Facebook channel on the Roku. Some users have reported an issue with this in the past, so here's a fix for that.

1. Log into your Facebook on a PC or MAC computer.
2. Click on the settings "gear" seen in the top right corner of your internet browser.
3. Select "Privacy Settings" option.
4. Click on the Apps icon you'll see (a small cube).
5. Click the "X" next to edit for the Roku App. A pop-up box will appear asking if you want to "Remove Roku?" app from your Facebook.
6. Click on "Remove" button and it will remove the app from your Facebook.

Now, go to facebook.com/device and follow the instructions there to link your Facebook account on your Roku. This should help link the account to your device.

Note: the above steps can also be used if you want to unlink your Facebook account from your Roku.

Unlink Device & Factory Reset

There may be some situations where you want to unlink your Roku device from your online account and perform a complete factory reset. This can be especially helpful if you are going to give someone else an old Roku, or sell it to someone.

1. On a PC/MC go to your account and log-in.
2. Go to "My Account" section.
3. Scroll down towards bottom of page to where your device is showing with a serial number.
4. Click on "Unlink." You'll be prompted with a pop-up box to click OK on before proceeding. The Roku device will be unlinked from your account.

Next, on your Roku device:

1. Go to "Settings" and select "Factory reset."
2. You'll receive a warning alerting you to what will happen.
3. Select "Factory reset" to begin the process to restore the device to its original settings.

The Roku will reset completely and reboot to the setup process. At this point, keep in mind you may need to pair the remote with the device again. To do this, simply open the battery compartment and hold the "pairing" button down for about 3 seconds until the lights come up on the remote. The remote should now control the device.

Roku Channels and Streaming Media Services

Part of the allure of the Roku electronic streaming device is the amount of channels available for your entertainment. No matter your personality or viewing preferences Roku has everything you could want, without the expense of a cable bill plaguing your expenses each month. While there are thousands of channels available, here is a look inside some of the popular Roku channels from the top categories, in an abbreviated form, of course.

Roku Channels

There are so many Roku channels to choose from. The following sections take a look at several different categories.

News

If you want to keep up with what is going on in the world, Roku News channels will provide you with access to all of the headlines you need at a moment's notice. There are 53 news and weather channels available on Roku in the Channel Store, including the following:

- Fox News
- NBC News
- CBS News
- Roku Newscaster
- Wall Street Journal Live
- Sky News
- Huff Post Live
- Free Speech TV
- Local News Outlets (geographical areas vary)

Family

Roku also provides family entertainment at its best. The Channel Store offers 43 Kids & Family channels like the following:

- PBS Kids
- Disney Channel
- Disney Junior
- National Geographic Kids
- Family TV
- CraftSmart
- Baby First
- Classic Cartoons

Sports

For the sports lover in your home, there are a whopping 64 sports channels available in the Roku Channel Store, including the likes of:

- WatchESPN
- CBS Sports
- XOS College Sports
- ESPN Radio

There is also popular programming available that revolves around a single sport, including basketball, hockey, tennis, MMA fighting, motor racing, women's sports and sailing.

International

For the international soul that lurks within, there are 35 channels that appeal to the entertainment values worldwide. Some of the Channel Store offerings in this genre include:

- Dish World Live International TV
- Kdrama
- DramaFever
- Tagesschau.de
- Brazilian TV
- Haiti TV Network
- EU Live
- Global TV Network
- African Info Media
- Arabic TV
- Mandarin TV
- Georgian Public Broadcaster

Additional Channel Categories

Additional channel categories in the Channel Store for your entertainment include:

- Games
- Science & Technology
- Food
- Fitness & Outdoor
- Special Interest
- Travel
- Religion & Spirituality

Streaming Media Services

Streaming Media Services are considered as subscription services that you must pay for, and access with each individual account's username and password. If you are a member, or wish to become a member of the following channels, here is what you can expect from the various choices available for streaming on Roku.

WWE Network

The WWE Network is a brand new platform that officially launched for Roku on Monday, February 24, 2014. This service, which costs $9.99 per month as of this publication, allows fans to access a wide library of old and new video content including past pay-per-views, highlight shows and new exclusive programs. To access the service, viewers will need to sign up for an account for the WWE Network at WWE.com website.

To add the channel on the Roku:

1. Go to the Channel Store and browse for "WWE."
2. Select the WWE Network channel and click to "Add Channel."
3. Click on "Yes" to confirm installation and the channel will be installed.

You'll need to go to the WWE Network channel on your Roku and enter your subscription login details to access the live streaming content on the channel.

See more details about the new 2014 service at http://www.wwe.com/help.

Netflix

Just as you would watch your favorite television shows and movies on Netflix on your laptop, tablet or smartphone, now you can stream the programming directly onto your television set when you have Roku.

Just as it is with all Netflix programming, you must be a subscriber to enjoy its content. And for $7.99 a month, you can be. Your Netflix subscription will give you complete on demand access to thousands of movies and television shows for your enjoyment. Thanks to Roku, you can now stream those movies and television shows directly to your larger screen, instead of counting on your laptop or tablet's smaller display. This is great for group viewing, or movie night with your sweetie. Finally, no one has to be responsible for holding the tablet steady, or burning their thighs with a laptop sitting on top of you for the duration of the programming. See more info at http://netflix.com.

Amazon Instant Video

Amazon Prime, or Prime Instant Video, is a media streaming service that is brought to you by the ever popular online giant amazon.com. This exciting service is available for $79 per year, and includes unlimited access to movies and television shows. The benefit of Amazon Prime is that its membership price comes with a free two day shipping upgrade on any item you order from Amazon, and access to thousands of eBooks.

The downfall is that its media streaming services include popular movies, but not always the newest releases. You can dip into the archives of some of your favorites, but if you are looking for the movies that are new to DVD each Tuesday, this service lacks that up to date availability. It does, however, bring you past television episodes and kid's programming at no extra charge, but will not provide you with the previous day's television programming for free. You can watch all of their offerings as often as you like, even if you want to watch the same episode from the first season of Downton Abbey for the thirteenth time. Just do not expect to see the current episodes for quite some time.

Hulu Plus

Hulu Plus, unlike Amazon Prime, will bring you an up to date programming, as soon as the following day. So if you missed your favorite Wednesday night sitcom, you can pick it up Thursday morning through Hulu Plus's media streaming services.

For a mere $7.99 Hulu Plus users can enjoy their favorite network offerings almost immediately, while having access to over 2000 movies that are in their database, and extended trailers for brand new movies that are still in the theaters. The main disadvantage of Hulu Plus is that you are still subjected to commercials, even though you are paying for the service, no matter how "limited" you are eluded to believe through services description. More info at Hulu.com website.

Crackle

Crackle provides a different take on movies and television series, providing full length, uncut and unfiltered content to the masses. You can choose to watch a block of Seinfeld episodes, or watch an old Bruce Lee movie at the drop of a hat. Crackle also provides original programming options, but no matter what you choose it is always free. Simply download the channel, register and setup a profile and get started. You can create a watch list, restrict access to minors and even share your findings, episodes or movies with friends and family members. See Crackle.com for more info.

Vudu

Vudu is a different from the other media streaming services in its entirety because it does not have a monthly membership fee. With over 15,000 movies in its library, and nearly 2000 of them in high definition, Vudu boasts the largest HD availability of its kind. The fees come into play, not through a monthly cost, but through rental and purchase fees. You can rent a movie on Vudu for two nights for two bucks. Not bad, actually. The problem is, if you want to view your favorite television programming, you are going to have to wait until the season is complete and it is offered on DVD -- and then you have to purchase the entire season. The pricing, starting at around $12.00 per season, is not the downfall -- especially if you are trying to catch up on the first couple of seasons of a new favorite. However, episode by episode purchases are non-existent for the time being. Head over to Vudu.com for more info.

HBO Go

HBO Go is the equivalent of having HBO on Demand, so you can keep up with the over 1400 programs they display, when you are ready to watch them. HBO Go provides access to every episode of every season of the best of HBO, all FREE with your HBO subscription. The bad news is, if you do not have cable or satellite, you will not have access to an HBO account. If you do not have an HBO account, you cannot access HBO Go.

If you have an HBO account, you can create a customized watchlist and catch up on your favorite HBO shows and hit movies at your convenience. You can also watch new episodes of your favorite shows and hit movies simultaneously as they premiere on HBO. You can also set a Series Pass to automatically send new episodes of your favorite HBO shows to your watchlist, and password protect your account to keep young viewers from watching explicit content when you are not around. HBO Go is free, and is included with the price of your HBO subscription. You will need to create a username and password to access Go on your Roku. Get more info at the HBOGo.com website.

MGo

If you do not have an existing cable or satellite service, MGo is for you. MGo allows you to watch the very latest Hollywood releases as a new pay-as-you-go digital movie and TV streaming service that makes it easier than ever to browse, purchase and enjoy the newest and best selection of movies and TV shows.

MGO offers the freshest new digital titles months and sometimes even years ahead of Netflix. Since you pay as you go, and the price varies based on what you are renting or buying, you do not have a pay a monthly subscription for the service, but simply for the content you want to view. MGo.com.

Music on Roku

Roku allows you to listen to all of your favorite music from the comfort of your very own home, and through the television, so it emanates around the room with ease while you clean or entertain.

Spotify

The Spotify channel allows users to stream music directly from their television while examining their available interests through song titles and artists or album titles and record labels, and even different genres. All Spotify users must have a Facebook account, and are required to login to the music service using that account to enjoy their offerings.

When you sign up for Spotify account you will receive a free six month subscription, and unlimited usability during that time. At the end of the trial period, song play is limited to only ten hours per month that is unveiled in two and one half hour increments per week, without a formal subscription. While using the ten free hours, listeners will have to contend with advertisements as a form of payment for the service. A subscription can be purchased that will give listeners access to music whenever they choose, without the ads. A premium service can also be purchased that will allow the music to stream at a higher bit rate, and offline access. (http://spotify.com)

Pandora

Pandora Radio is a music channel that allows listeners to search for musical selections by entering song titles or artists they love, and receiving a radio "Station" as a result. If the user enters "Brad Paisley" as a search artist, it will also play artists like Toby Keith and Kenny Chesney, and provide those musical selections as a "Station". So if someone says they are listening to Kenny Chesney radio on Pandora, it is simply the result of entering his name, and receiving music that relates to the genre and sound. Pandora also provides a feedback system that allows users to give positive and negative evaluations of the musical offerings, to hone their specific listening experience.

Listeners can purchase songs or albums too, but absolutely do not have to in order to enjoy the service. There are two subscription options available with the Pandora channel First: Free service, which is supplied through advertisements, or Second: A no-ad fee-based subscription. The Pandora library is comprised of 800,000 tracks, from over 80,000 artists.

MOG

MOG provides users with access to over ten millions songs and over one million albums on demand, which allows you to listen to whatever you want, whenever you want. What's more is that MOG will compile your listening habits to deliver a personalized listening experience that includes all of your favorite genres and music styles, so you never have to search for what you love to hear.

Amazon Cloud Player

The Amazon Cloud Player on your Roku works exactly like the one on your laptop or tablet, and allows you to enjoy your music everywhere. You can securely store your music in the cloud, and then stream it to your Roku player. Keep in mind, every time you purchase MP3s from Amazon, they are automatically added to Cloud Player for free!

TuneIn Radio

TuneIn Radio provides an international approach to listening to music, sports, news and entertainment offerings from around the world. You can choose from live global newscasts to international radio stations and specialty podcasts that you cannot get anywhere else. So keep your ears open the next time you travel abroad, so you know what to look for on your TuneIn Radio when you return home.

Live365

Live365 goes one step further in providing you with lyrical and musical entertainment, as it delivers over 7000 radio stations directly to your Roku. This means you can literally swim in a sea of stations to get the most listening pleasure that is available for your personal enjoyment. Likewise, family members who may not be listening to newer music simply because they are tied to their iPods or MP3 players downloads can enjoy an eye opening experience with these thousands of stations just waiting to be heard.

Vevo

Vevo goes a step further than offering radio station-style music, as it literally brings your television to life with full length music videos! Watch your favorite music videos and discover new ones on the VEVO Roku channel. This channel provides access to VEVO's entire catalog of 75,000 music videos from more than 21,000 artists. Who needs MTV or VH1 when you have VEVO at your fingertips? The best part is you are not subjected to videos in the order the station wants to play them. You can physically pick and choose your favorites to watch over and over again, or display new versions you have yet to see. You can even invite your friends over to share in the mystery of what was once one of the best ways to enjoy music: Through music videos!

Roku Private Channels to Increase your Viewing Opportunities

Private channels are generally hidden from the Roku channel store because some of them are in beta form, or require special subscriptions, or feature adult content that Roku doesn't want as part of the general store. I've assembled a list of some of the better private channels available for your Roku below.

Keep in mind some of these channels may not work on specific older model Rokus or in certain regions. Some may require additional fees or sign-ups to use, but for the most part there are some great free channels in the list below. To add the channels, use the instructions from earlier in this book or check out this helpful article with a video <http://www.techmediasource.com/roku-private-channels-how-to-add-hidden-channels/> on Roku private channels.

NOTE: *Many hidden channels are not officially approved by Roku, so it is best to be very cautious as far as providing any personal or other information that is requested.*

Top Private Channels List

America's Television Network (Code: americastv) – A free channel which provides network, live and 24/7 content in hi-definition. Includes movies, TV shows, music, games and other streaming content.

Comedy Fun Flix (Code: comedyfunflix) – A selection various low budget comedy movies for free.

CLASSIC-TV (Code: PEFL13) – As of this publication, this private channel appears to only be working on ROKU 1 and 2 devices, but not the ROKU 3. It features various entertainment variety channels including a Global Travel Guide, movie trailers, MusicJukeBoxTV, 70s Rewind TV and more.

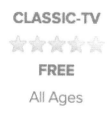

CLASSIC-TV

FREE

All Ages

Force One Network Radio (Code: A5GGU) – Provides the latest music from genres including hip hop, R&B, electro, dance, club, disco and more. This channel features various guest DJ sets and mastermixes with nothing but the top classics in music streaming to your Roku.

iTunes Podcasts (Code: ITPC) – Use this channel to watch hundreds of different video feeds in the top podcasts, genres or providers categories. Podcast genres include arts, business, comedy, sports, technology, society, health and much more!

Nowhere TV (Code: H9DWC) – This private channel gives all sorts of sports and news stream updates including but not limited to live local news, sports, public radio, government, home and garden, international and other great streaming content.
Note: There are lots of live TV channels under the "International" category for this private Roku channel.

Nowhere Music (code: nowheremusic) – A google client, which requires you to have an account set up, in order to stream.

PlayOn (Code: MYPLAYON) – This is certainly one of the most popular private channels for Roku, as it allows you to watch all sorts of videos, TV shows and movies by streaming content from your PC to the Roku. It may require some extra work to get this one set up via your PC by installing the "PlayOn media server," but once it is there, you may find yourself loving the selection that this private channel offers!
Note: This is a year or lifetime-subscription-based channel, with more info at http://www.playon.tv.

Rokagram Beta (Code: rokagrambeta) – Provides an "easy-to-use" Instagram channel for the Roku. With this private channel you can view images, follow friends and check out the latest trends.

Television Shopping (Code: TVS) – This free channel provides streams of QVC, QVC UK, Shop NBC & Jewelry Television feeds.

Welcome to Hip Hop (Code: 2KC5S) –The streaming hip-hop music station, provides viewers 24/7 enjoyment, of streaming songs and videos, both past and present in the world of hip-hop.

LibriVox (Code: Q4DSY) – LibriVox offers free audiobooks. These are audio readings from books that are free in the public domain.

NASATV (HD) (Code: ENDLESSNASA) – NASA TV streaming straight to your Roku device. Interesting information about what is going on at NASA.

The Jack and Holly Channel (Code: jackandholly) – Jack and Holly are two of the most lovable pre-school characters in the UK. The characters aim to inform, educate, and entertain pre-schoolers and their parents by talking about key life questions. The episodes offer songs, great animations, and a few tall stories. Plenty of fun entertainment for pre-school aged children.

Force One Network Radio (Code: A5GGU) – This station provides listeners with a rebirth of old, classic, R&B, and hip-hop tunes, which are no longer played on mainstream radio and TV today.

Apply Makeup Tips (Code: makeuptips) – A channel providing free makeup tips, to help women achieve flawless design features, and distinct new styles.

Holy Spirit & Fire TV (code: worldnetchrist) – A free station, providing streaming religious services and sermons.

Joaquin FX Trading (code: TFEX6) – A station for traders in the varying markets, providing information on the latest trends, and news in the trading world.

Grid Free Survival (code: gridfreesurvival) – A free streaming station, which teaches you how to live off the grid, and eliminate your footprint.

JellyTellyFlat2 (code: ftflat2) – A children's Christian video station, providing over 100 streaming hours of video.

Avian Post (code: avianpost) – This site provides short form bird watching documentaries, for outdoor enthusiasts.

Gem City Sports (code: gemcitysports) – A station for Dayton sports fans, providing streaming sports commentary and news.

V Network (Code: vnetwork) – A streaming station providing shows ranging from entertainment to news, to real estate in the local area.

Volume TV (code: volumetv) – An internet TV station, featuring the latest artists, and their new sounds.

Qvivo Beta (code: qvivobeta) – Build your social media presence on the cloud, streaming through this station.

Rock Bottom Radio (code: rockbottom) – Online streaming rock and roll, and heavy metal, from a variety of artists.

Hot Seat with Wally George (code: wally) – A streaming station providing viewers with a revival, of this classic 80s show.

Guitarded Radio (code: gtrded) – An online streaming station, which celebrates the guitar. It celebrates the many forms, as well as the different genres of music the guitar is used in.

SDCT (code: sdct2013) – A live, streaming station, which showcases a number of independent artists, film, and classic TV.

Maximum Threshold TV (code: maximumthresholdTV) – Commentary, news, and upcoming events, in the world of rock, and metal. The station also provides the latest news and events, for the genre and artists.

Next News Network (code: EWGGU) – Features breaking news, live updates, and exclusive news and reports, which may have an effect on your freedoms and liberties.

Inside Nano Tech (code: insidentek) – Station featuring the latest news and reports, about the Nano.

Aliento Vision (code: aliento) – Hispanic streaming station, featuring a variety of news and updates.

Jhanjar TV Live (code: jhanjar) – Worldwide S Asian news and updates, 24 hours a day.

Hmong USA TV (code: hmong) – A US station, which telecasts in the Hmong language as well as other SE Asian languages.

On the Mike (code: 8AESS) – A celebrity driven music and radio show, which is one of the most popular in the US.

Jesus live Network (code: jesuslivenetwork) – Ministries, music, documentaries, and sermons, all in one place.

VP Live (code: vapeteam) – All the news and updates you want to know about electronic cigarettes; from safety, to health related issues.

Kim Clement TV (code: kctv) – Christian singing, sermons, as well as piano lessons.

Senior Care TV (code: pucdz) – Information and updates on senior care, living facilities, and home care.

Post it news beta (code: postitnewsbeta) – Beta app channel, for news on the latest updates.

Blog your wine (code: blogyourwine) – A streaming station about wine, and information on new wine development methods.

MSdevtv (code: msdevtv) – A station which may be used by Microsoft developers, with video and streaming news.

Game View TV (code: gvtv) – News for Oklahoma City high school sports, games, and upcoming local events.

Church Pond (code: churchpond) – A station for Christians to watch and listen to the latest teachings and sermons.

Vtnbeta (code: vtnbeta) – A streaming source for Christian programming.

360 Music Television (code: 360beta) – A streaming music station, for the latest music, videos, and artists, all in one place.

Citrus Daily (code: citrusdaily) – Daily news, sports, and information, for the space coast, and Cape Canaveral area, in FL.

JoCo Roku (code: joco) – A Johnathan Coulton music player, where fans can stream all of the tracks that he has created.

Map channel beta (code: map) – A streaming station which allows you to scroll down maps from around the solar system, and allows you to get down as close as street view.

Motorcycles Skateboarding (code: skate) – A channel for motorcycles and skateboard vlogging.

Ahlul Bayat TV (code: abtv1) – The first free station from an Islamic provide, to stream in exclusively in English TV.

tv3q (code: tv3q) – A station which streams over 50 channels from Hong Kong, Taiwan and Japan.

Israel Live (code: islive) – This station broadcasts direct from Israel, from two independent stations.

Westside Church Omaha (code: westside) – A streaming Omaha station, providing sermons from the church directly.

Media for Christ (code: mfc) – A station providing the word of God, and gospel preaching.

Threshold (code: thresholdplus) – For the sci-fi fan; horror, fantasy, and sci-fi thriller movies and TV shows, are broadcast on this station.

Bloomberg TV+ (code: btvplus) – A streaming station, which allows you to stream directly from any of the 5 available stations for trading and financial news.

Telemadrid (code: Madrid) – A streaming station which allows fans, or locals from Madrid, to view the latest news, sports, and information that comes directly from the country, streamed on to your Roku device.

Rightway TV (code: rightway) – A station streaming news from the top right minded voices today.

There are hundreds of Roku hidden channels to choose from. These are a few of the top choices, some free, and some paid for.

Roku Accessories

As with any great electronic device, accessories are not that far away! Roku is no exception, and has a number of accessories that will enhance your entertainment enjoyment.

The Remotes

One has to think that television would not be as popular as it is today if the remote were never invented. Thank goodness for small favors, no?

The Roku has a number of remote control options available to the masses, proving that point to be exact.

Roku 3 Enhanced Game Remote

The Roku 3 Enhanced Game Remote comes complete with a built-in headphone jack, so you can plug your headphones directly into the remote and enjoy a private listening session. It comes with motion-control and gaming buttons for hours of gaming fun. This state of the art remote allows you to control the Roku box behind closed cabinets and walls. This version works only with the Roku 3, and is $24.99.

Roku 2 Enhanced Remote

The Roku 2 Enhanced Remote features a built-in headphone jack for private listening, and allows you to navigate movies, music, shows and more. RF technology gives you control of your Roku player behind closed cabinets and walls. This version works only with the Roku 2 or Roku 3, and is $24.99.

Roku Standard Remote with Channel Shortcut Buttons

The infrared Standard Remote controls movies, music, shows and more, and includes channel shortcut buttons for access to all of your favorites (Netflix, M-GO, Amazon Instant Video and Blockbuster on Demand) with one touch of a button. This version works with all Roku models, except the Roku Streaming Stick and is available for $14.99.

Roku Standard Remote

Our standard remote puts you in the driver's seat to control movies, music, shows and more using infrared (IR) technology. This version works with all Roku models except the Roku Streaming Stick and is available for $14.99.

Roku Streaming Stick Enhanced Remote

Have a Roku Streaming Stick plugged into a Roku Ready device? The Roku Streaming Stick enhanced remote controls your Roku interface and includes motion control for gaming. The remote uses Wi-Fi Direct technology, and requires a Roku Ready device and Roku Streaming Stick to operate. This remote is available for $24.99.

Roku 2 Enhanced Game Remote Pack

Turn your Roku 2 into a great place for the coolest games with this Roku 2 enhanced game remote with motion control and one 2 GB MicroSD card to boost your game storage capacity. This accessory uses Bluetooth technology, and works only with the Roku 2 XS and Roku 2 XD. The package is available for $35.

Roku Premium In-ear Headphones Family Pack

Keep the peace in the house with the 4-pack of premium sound in-ear headphones that plug right into the Roku 3 and Roku 2 remotes with built-in headphone jack. These headphones come with three sets of ear tips for the perfect fit, and are available for $39.99. You can also purchase a single pair of premium sound headphones for $14.99, or a standard (non-premium) pair for $7.99.

Standard and Replacement Power Supplies

No matter which model of Roku you have, you can buy a replacement power supply cord for $9.99. Just be sure to choose the right version for your outlet.

Six Foot HDMI Cable

Adding a lengthy, easy-to-use HDMI cable will help deliver top HD video and digital audio quality to your Roku 3, Roku 2 or Roku 1 for a mere $9.99.

Two Gigabyte MicroSD Card

This tiny MicroSD card adds extra storage capacity for hundreds more channels and games to your Roku 3, and is available from Roku for a mere $4.99.

USB flash drives

Several Roku devices have the option to plug in a USB jump/flash drive or a USB external drive. This is a great way to get a lot extra media onto your device such as your music collection, videos and photos.

There are 128GB jump/flash drives currently available for $50 or under.

In addition, there are 500GB external plug-in hard drives available for just over $50. It's all really a matter of how much extra content you want to be able to use on your device.

Slingbox and SlingPlayer

The Slingbox is an audio-video device that allows you to watch what is on your TV anywhere you go. You simply connect the Slingbox to your TV's set-top box, your television, and your home network. Then you can access TV programs and even control or program your DVR while away from home. It has been made easier for Roku owners to do this with a SlingPlayer for Roku app on iPhone or Android phones. You can use the apps to stream your home Slingbox TV programs to your Roku anywhere you bring it and connect to a TV!

Current models include the Sling Media Slingbox 350 and the Sling Media Slingbox 500. See more information at Slingbox.com website.

Additional Roku Help & Support

You can get additional help at Roku's official website, by looking at your account options. Among them are options to view your account information including name, log-in and shipping address. You can also set a PIN preference for when making purchases on your Roku device. This can help prevent accidental orders from being made on your device.

https://owner.roku.com/

On your account page, you can also perform the following:

- Add or update your payment method on file.
- Add private channels.
- View your purchase history.
- Manage your various channel store subscriptions.
- Deactivate your account.

In addition, there is the option to link, rename or unlink Roku devices to your online account.

There is also Roku Support available with "Featured Questions" that users can look through to get inside info such as new features, updates and channels that are on the way for the device.

http://support.roku.com/home

Another great resource for extra tips, tricks and help is the Roku Forums website. The forums have lots of other Roku users providing each other with troubleshooting solutions, as well as Roku personnel answering questions about new features and aspects of using the device. See the Roku Forums at:

http://forums.roku.com/

Conclusion: Roku Beyond TV

With the help of Roku TV, you may even be able to cut the cable cord once and for all. For more information about cutting the cord completely check out my book about ditching cable once and for all. With the right setup for you, you will never even miss it.

However, even if you subscribe to cable, Roku can help you get TV in rooms where you do not have cable installed. You can also access your premium channels the HBO using HBO Go in rooms where you do not have a cable box.

Roku is also a fantastic way to stream videos from YouTube or movies from Netflix or Redbox Instant Access, or any number of moving streaming services. In addition, Roku offers the most fitness channels of any streaming device, so you can get in your workout at home without things becoming stale since there are so many options available.

There are so many private channels available for Roku. I have just listed a few of my favorites that I have on my own device. However, you can explore these many options, and find all sorts of interesting content to suit your style, preferences, and desires. There truly is something for everyone on Roku.

The great thing is that Roku allows you to go beyond just TV. You can listen to music, watch videos, or play slideshows of your own content using Roku depending on what services you choose to use or subscribe to. The sky is the limit with this handy little device, and there are so many different options, that there is sure to be one to fit your lifestyle and needs perfectly. You can even take Roku with you on the road with the Roku 3400M Streaming Stick.

Isn't it amazing to have options? Roku gives you an unbelievable amount of options, and you can watch it in addition to your cable, or use it to escape cable completely! The choice is yours.

More Books by Shelby Johnson

iPad Mini User's Guide: Simple Tips and Tricks to Unleash the Power of your Tablet!

iPhone 5 (5C & 5S) User's Manual: Tips and Tricks to Unleash the Power of Your Smartphone! (includes iOS 7)

Kindle Fire HDX & HD User's Guide Book: Unleash the Power of Your Tablet!

Facebook for Beginners: Navigating the Social Network

Kindle Paperwhite User's Manual: Guide to Enjoying your E-reader!

How to Get Rid of Cable TV & Save Money: Watch Digital TV & Live Stream Online Media

Chromecast User Manual: Guide to Stream to Your TV (w/Extra Tips & Tricks!)

Google Nexus 7 User's Manual: Tablet Guide Book with Tips & Tricks!

Printed in Great Britain
by Amazon.co.uk, Ltd.,
Marston Gate.